Chocolate

Or, An Indian Drinke By the wise and Moderate use whereof, Health is preserved, Sicknesse D ted, and Cured, especially the Plague of arly called Th

Antonio Colmenero de Ledesma

Alpha Editions

This edition published in 2024

ISBN : 9789367244586

Design and Setting By
Alpha Editions
www.alphaedis.com
Email - info@alphaedis.com

As per information held with us this book is in Public Domain.
This book is a reproduction of an important historical work. Alpha Editions uses the best technology to reproduce historical work in the same manner it was first published to preserve its original nature. Any marks or number seen are left intentionally to preserve its true form.

THE TRANSLATOR,
To every Individuall Man, and Woman, Learn'd, or unlearn'd, Honest, or Dishonest: In the due Praise of Divine *CHOCOLATE*.

Doctors *lay by your* Irksome Books
And all ye Petty-Fogging Rookes
Leave Quacking; *and* Enucleate
The vertues *of our* Chocolate.

Let th' Universall Medicine
(Made up of Dead-mens Bones *and* Skin,*)*
Be henceforth Illegitimate,
And yeild to Soveraigne-Chocolate.

Let Bawdy-Baths *be us'd no more;*
Nor Smoaky-Stoves *but by the whore*
Of Babilon: *since* Happy-Fate
Hath Blessed *us with* Chocolate.

Let old Punctæus *Greaze his* shooes
With his Mock-Balsome: *and Abuse*
No more the World: But Meditate
The Excellence *of* Chocolate.

Let Doctor Trigg *(who so Excells)*
No longer Trudge to Westwood-Wells:
For though that water Expurgate,
'Tis but the Dreggs *of* Chocolate.

Let all the Paracelsian *Crew*
Who can Extract Christian *from* Jew;
Or out of Monarchy, *A* State,
Breake àll their Stills *for* Chocolate.

Tell us no more of Weapon-Salve,
But rather Doome us to a Grave:
For sure our wounds will Ulcerate,
Unlesse they're wash'd *with* Chocolate.

The Thriving Saint, *who will not come*
Within a Sack-Shop's *Bowzing-Roome*

(His Spirit *to* Exhilerate*)*
Drinkes Bowles *(at home) of* Chocolate.

His Spouse *when she (*Brimfull *of* Sense*)*
Doth want her due Benevolence,
And Babes *of* Grace *would* Propagate,
Is alwayes Sipping Chocolate.

The Roaring-Crew *of* Gallant-Ones
Whose Marrow *Rotts within their* Bones*:*
Their Bodyes *quickly* Regulate,
If once but Sous'd *in* Chocolate.

Young Heires *that have more* Land *then Wit,*
When once they doe but Tast *of it,*
Will rather spend their whole Estate,
Then weaned *be from* Chocolate.

The Nut-Browne-Lasses *of the Land*
Whom Nature *vayl'd in* Face *and* Hand,
Are quickly Beauties *of* High-Rate,
By one small Draught *of* Chocolate.

Besides, it saves the Moneys *lost*
Each day in Patches, *which did cost*
Them deare, untill of Late
They found this Heavenly Chocolate.

Nor need the Women *longer* grieve
Who spend *their* Oyle, *yet not* conceive,
For 'tis a Helpe-Immediate,
If such but Lick *of* Chocolate.

Consumptions *too (be well assur'd)*
Are no lesse soone *then* soundly *cur'd:*
(Excepting such as doe Relate
Unto the Purse*) by* Chocolate.

Nay more: It's vertue *is so much,*
That if a Lady *get a* Touch,
Her griefe it will Extenuate,
If she but smell *of* Chocolate.

The Feeble-Man, *whom* Nature *Tyes*
To doe his Mistresse's Drudgeries*;*
O how it will his minde Elate,
If shee *allow him* Chocolate*!*

'Twill make Old women Young *and* Fresh;
Create New-Motions *of the* Flesh,
And cause them long for you know what,
If they but Tast *of* Chocolate.

There's ne're a Common Counsell-Man,
Whose Life *would Reach unto a* Span,
Should he not Well-Affect *the* State,
And First *and* Last *Drinke* Chocolate.

Nor e're a Citizen*'s Chast wife,*
That ever shall prolong her Life,
(Whilst open *stands* Her Posterne-Gate*)*
Unlesse she drinke *of* Chocolate.

Nor dost the Levite *any Harme,*
It keepeth his Devotion *warme,*
And eke the Hayre *upon his* Pate,
So long as he drinkes Chocolate.

Both High *and* Low, *both* Rich *and* Poore
My Lord, *my* Lady, *and his* ——
With all the Folkes *at* Billingsgate,
Bow, Bow *your* Hamms *to* Chocolate.

Don Diego de Vadesforte.

To the Author,

Great Don, Grandee of *Spaine*, Illostrissimo of *Venice*, High and mighty King of *Candie*, Great Bashaw of *Babilon*, Prince of the Moone, Lord of the Seven Starres, Governour of the Castle of *Comfort*, Sole Admirall of the Floating *Caravan*, Author of Th' *Europian* Mercury, Chiefe Generall and Admirall of the Invisible Fleet and Army of *Terra Incognita*,

Cap. JAMES WADSWORTH.

The Allowance of Melchor De *Lara*, Physitian Generall for the Kingdome of *Spaine*.

I Doctor *Melchor de Lara* Physitian Generall for the Kingdom of *Spaine*, at the command of *Don John de Velasco*, and *Asebedo*, Vicar Generall of *Madrid*, have seene this Treatise of *Chocolate*, composed by *Antonio Colmenero* of *Ledesma*; which is very learned, and curious, and therefore it ought to be Licensed for the Presse; it containing nothing contrary to good manners; and cannot but be very pleasing to those, who are affected to *Chocolate*. In testimony whereof, I have subscribed my Name, in *Madrid* the 23. day of *August*. 1631.

Melchor de Lara.

The Testimoniall of John de *Mena*, Doctor and Physitian to the King of *Spaine*.

I John de Mena, *Physitian to his Majesty, and one of the Counsell Generall of the Inquisition, have seene this Treatise of* Chocolate *(composed by Doctor* Antonio Colmenero *of* Ledesma*) by command of the* Supreame Royall Court of Justice*: which containeth nothing contrary to good Manners, and the Subject if very learnedly handled, and with great Iudgement; and no doubt, but it will give much pleasure and content to all those, who are affected to* Chocolate*; and therefore may be printed: And in confirmation of this truth, I have hereto subscribed my Name the* 17. *of* Septemb. 1631.

John de Mena *Doctor in Physicke.*

To the Reader.

The number is so great of those, who, in these times, drinke *Chocolate*, that not only in the *Indies*, where this kind of Drink hath its originall; but it is also much used in *Spain*, *Italy* and *Flanders*, and particularly at the Cour. And many doe speake diversly of it, according to the benefit, or hurt, they receive from it: Some saying, that it is stopping: Others, and those the greater part, that it makes one fat: Others, that the use of it strengthens the stomacke: Others, that it heates, and burns them: And others say, that although they take it every houre, and in the Dogdayes, yet they finde themselves well with it. And therefore my desire is, to take this paines, for the pleasure, and profit of the publicke; endeavouring to accommodate it to the content of all, according to the variety of those things, wherewith it may be mixt; that so every man may make choise of that, which shal be most agreeable to his disposition. I have not seene any, who hath written any thing, concerning this drinke; but onely a Physitian of *Marchena*, who (as it seemes) writ onely by Relation; holding an opinion, that the *Chocolate* is stopping, because that *Cacao* (the principall Ingredient of which it is made) is cold, and dry. But because this onely reason, may not have power to keepe some from the use of it, who are troubled with Opilations; I thinke fit to defend this *Confection*, with Philosophicall Reasons, against any whosoever will condemne this Drinke, which is so wholesome, and so good, knowing how to make the Paste in that manner, that it may be agreeable to divers dispositions, in the moderate drinking of it. And so, with all possible brevity, shall distinguish and divide this Treatise into foure poynts, or Heads. In the first place I shall declare, what *Chocolate* is; and what are the Qualities of *Cacao*, and the other Ingredients of this *Confection*; where I shall treat of the Receipt set downe by the aforesaid Author of *Marchena*, and declare my opinion concerning the same. The second point shall treat of the Quality, which resulteth out of the mixture of these Simples, which are put into it. In the third place the manner of Compounding; and how many wayes they use to drink it in the *Indies*. In the fourth, and last place I shall treat of the Quantity; and how it ought to be taken; at what time; and by what persons.

The first Point.

Concerning the first Point, I say, that *Chocolate* is a name of the *Indians*; which in our vulgar Castilian, we may call a certaine *Confection*, in which (among the Ingredients) the principall *Basis*, and Foundation, is the *Cacao*; of whose Nature and Quality it is necessary first to treat: And therefore I say, according to the common received opinion, that it is cold, and dry, *à prædominio*; that is to say, that though it be true, that every Simple containes in it the Qualities of the foure Elements, in the action, and re-action, which it hath in it, yet there results another distinct quality, which we call Complexion.

This Quality or Complexion, which ariseth of this Mixture, is not alwayes one, and the same; neither hath it the effect in all the mixtures, but they may be varied nine wayes; four *Simple*, from whence one onely quality doth abound; and foure *Compounded*, from whence two Symbolizing qualities are predominant; and one other, which we call *ad pondus*, which is of all these fore-said qualities, which are in *æquilibrio*, that is to say, in equall measure and degree.

Of all these the Complexion of *Cacao* is composed, since there arise two qualities, which are cold, and dry; and in the substance, that rules them, hath it *restringent* and *obstructive*, of the nature of the Element of the *Earth*. And then, as it is a Mixed, and not a simple Element, it must needs have parts correspondent to the rest of the Elements; and particularly, it partakees (and that, not a little) of those, which correspond with the Element of Aire, that is, Heat and Moysture, which are governed by the Unctious parts; there being drawne out of the *Cacao* much Butter, which, in the *Indies* I have seene drawne out if it, for the Face, by the *Criollas*.

It may Philosophically be objected, in this manner: *Two contrary Qualities, and Disagreeing, cannot be* in gradu intenso, *in one and the same Subject:* Cacao *is cold and drie, in predominency: Therefore, it cannot have the qualities contrary to those; which are Heat, and Moysture. The first Proposition is most certaine, and grounded upon good Philosophy: The second is consented unto, by all: The third, which is the Conclusion, is regular.*

It cannot be denyed, but that the *Argument* is very strong, and these reasons being considered by him of *Marchena*, have made him affirme, that *Chocolate* is Obstructive; it seeming to be contrary to Philosophy, that in it there should be found *Heat* and *Moysture, in gradu intenso*; and to be so likewise in *Cold* and *Dry*.

To this, there are two things to be answered: One, that he never saw the experience of drawing out the Butter, which I have done; and that when the *Chocolate* is made without adding any thing to the dryed Powder, which is incorporated, onely by beating it well together, and is united, and made into a Paste, which is a signe, that there is a moist, and glutinous part, which, of necessity, must correspond with the Element of Aire.

The other reason, we will draw from Philosophy; affirming that, in the *Cacao*, there are different substances. In the one, that is to say, in that, which is not so fat, it hath a greater quantity of the Oylie, then of the earthie Substance; and in the fatter part, it hath more of the earthy than of the Oily substance. In these there is Heate and Moysture in predominancy; and in the other, cold and dry.

Notwithstanding that it is hard to be believed, that in one and the same substance, and so little of the *Cacao*, it can have substances so different: To the end that it may appear more easie, clear, and evident, first we see it in the *Rubarbe*, which hath in it hot and soluble parts, and parts which are Binding, Cold and Dry, which have a vertue to strengthen, binde, and stop the loosenesse of the Belly: I say also, that he that sees and considers the steele, so much of the nature of the earth, as being heavy, thick, cold, and dry; it seemes to be thought unproper for the curing of Opilations, but rather to be apt to encrease them; and yet it is given for a proper remedy against them.

This difficulty is cleared thus, that though it be true, that it hath much of the Earthy part; yet it hath also parts of Sulphur, and of quick silver, which doe open, and disopilate; neither doth it so, untill it be helped by Art, as it is ground, stirred, and made fine, in the preparing of it; the Sulphurous parts, and those of quick-silver, being thinne, active, and penetrative, they mingle, at the last with those parts, which are Earthy and astringent: Insomuch, that they being mingled after this manner one with another, we cannot now say, that the steele is astringent, but rather, that it is penetrative, attenuating and opening. Let us prove this Doctrine by Authorities; and let the first be from *Gallen, l. 3.* of the qualities of Simples, *c. 14.* Where, first of all he teacheth, that almost all those Medicines, which, to our sence, seeme to be *Simple*, are notwithstanding naturally *Compounded*, containing in themselves contrary qualities; and that is to say, a quality to expell, and to retaine; to incrassate, and attenuate; to rarifie, and to condense. Neither are we to wonder at it, it being understood, that in every fore-said Medicine, there is a quality to heat, and to coole; to moisten and to dry. And whatsoever Medicine it be, it hath in it, thick, and thinne parts; rare, and dense; soft, and hard. And in the fifteenth Chapter following, in the same Book, he puts an example of the Broth of a Cock, which moves the Belly; and the flesh hath the vertue to bind. He puts also the example of the *Aloes*, which if it be washt, looseth the Purgative vertue; or that which it hath, is but weake.

That this differing vertue, and faculty, is found in divers substances, or parts of simple Medicaments, *Gallen* shewes in the first Booke of his simple Medicines, and the seventeenth Chapter, bringing the example of Milke; in which, three substances are found, and separated, that is to say, the substance of Cheese, which hath the vertue to stop the Fluxe of the Belly; and the substance of Whay, which is purging; and Butter, as it is expressed in the said *Gallen, Cap. 15.* Also we finde in Wine which is in the Must, three substances, that is to say, earth, which is the chiefe; and a thinner substance, which is the flower, and may be called the scum, or froath: and a third substance which we properly call Wine; And every one of these substances, containes in it selfe divers qualities, and vertues; in the colour, in the smell, and in other Accidents.

Aristotle in the fourth Book of the Meteors and the first Chapter, treating of Putrefaction, he found the same substances; and in the second Chapter next following, where he that is curious may read it. And also by the Doctrine of *Galen*, and of *Aristotle*, divers substances are attributed to every of the mixt under one and the same forme and quantity; which is very conformable to reason, if we consider, that every Aliment be it never so simple, begets, and produceth in the liver, foure humours, not onely differing in temper, but also in substance; and begets more or lesse of that humour, according as that Aliment hath more or fewer parts corresponding to the substance of that humour, which is most ingendred. And so in cold diseases, we give warme nourishment; and cold nourishment, in hot diseases.

From which evident examples, and many others, which we might produce to this purpose, we may gather, that, when we grind and stir the *Cacao*, the divers parts, which Nature hath given it, doe artificially, and intimately mixe themselves one with another; and so the unctuous, warme, and moist parts, mingled with the earthy (as we have said of the steele) represses, and leaves them not so binding, as they were before; but rather with a mediocritie, more inclining to the warme, and moist temper of the Aire, then to the cold and dry of the Earth; as it doth appear when it is made fit to drinke; that you scarce give it two turnes with the Molinet when there riseth a fatty scumme: by which you may see how much it partaketh of the Oylie part.

From which doctrine I gather, that the Author of *Marchena*, was in an errour, who, writing of *Chocolate*, saith that it causeth Opilations, because *Cacao* is astringent; as if that astriction were not corrected, by the intimate mixing of one part with another, by meanes of the grinding, as is said before. Besides, it having so many ingredients, which are naturally hot, it must of necessity have this effect; that is to say, to open, attenuate, and not to binde; and, indeed, there is no cause of bringing more examples, or producing more reasons, for this truth, then that which we see in the *Cacao* it self: which, if it be not stirred, and compounded, as aforesaid, to make the *Chocolate*. But eating of it, as it is in the fruite, as the *Criollas* eate it in the *Indies*, it doth notably obstruct, and cause stoppings; for no other cause but this, that the divers substances which it containes, are not perfectly mingled by the mastication onely, but require the artificiall mixture, which we have spoken of before.

Besides, our Adversary should have considered, and called to his memory, the first rudiments of Philosophy, that *à dicto secundum quid, ad dictum simpliciter, non valet consequentia*; As it is not enough to say, the Black-a-Moore is white, because his teeth are white; for he may be blacke, though he hath white teeth; and so it is not enough to say, that the *Cacao* is stopping; and therefore the Confection, which is made of it, is also stopping.

The Tree, which beares this fruit, is so delicate; and the earth, where it growes, is so extreme hot, that to keepe the tree from being consumed by the Sun, they first plant other trees; and when they are growne up to a good height, then they plant the *Cacao* trees; that when it first shewes it selfe above the ground, those trees which are already growne, may shelter it from the Sunne; and the fruit doth not grow naked, but ten or twelve of them are in one Gorde or Cod, which is of the bignesse of a greate black Figge, or bigger, and of the same forme, and colour.

There are two sorts of *Cacao*; the one is common, which is of a gray colour, inclining towards red; the other is broader and bigger, which they call *Patlaxte*, and this is white, and more drying; whereby it causeth watchfulnesse, and drives away sleepe, and therefore it is not so usefull, as the ordinary. This shall suffice to be said of the *Cacao*.

And as for the rest of the ingredients, which make our *Chocolaticall* Confection, there is notable variety; because some doe put into it black Pepper, and also *Tauasco*A red roote like madder.; which is not proper, because it is so hot and dry; but onely for one, who hath a very cold Liver. And of this opinion, was a certaine Doctor of the University of *Mexico*, of whom a Religious man of good credit told me, that he finding the ordinary round Pepper was not fit to bring his purpose about, and to the end, he might discover, whether the long red pepper were more proper, he made triall upon the liver of a Sheepe; and putting the ordinary pepper on one side, and the red pepperChile. on the other, after 24 hours, the part, where the ordinary pepper lay, was dryed up; and the other part continued moist, as if nothing had bin thrown upon it.

The Receipt of him who wrote at *Marchena*, is this: Of *Cacaos*, 700; of white Sugar, one pound and a halfe; Cinnamon, 2. ounces; of long red pepper, 14. of Cloves, halfe an ounce: Three Cods of the Logwood or Campeche tree; or in steade of that, the weight of 2. Reals, or a shilling of Anniseeds; as much of *Agiote*, as will give the colour, which is about the quantity of a Hasell-nut. Some put in Almons, kernells of Nuts, and Orenge-flower-water.

Concerning this Receipt I shall first say, This shooe will not fit every foote; but for those, who have diseases, or are inclining to be infirme, you may either adde, or take away, according to the necessity, and temperature of every one: and I hold it not amisse, that Sugar be put into it, when it is drunke, so that it be according to the quantity I shall hereafter set downe. And sometimes they make Tablets of the Sugar, and the *Chocolate* together: which they doe onely to please the Pallats, as the Dames of *Mexico* doe use it; and they are there sold in shops, and are confected and eaten like other sweet-meats. For the Cloves, which are put into this drinke, by the Author aforesaid, the best Writers of this Composition use them not; peradventure upon this reason: that although they take away the ill savour of the mouth, they binde; as a learned Writer hath exprest in these verses:

> Fœtorem emendat oris Cariophilia fœdum;
> Constringunt ventrem, primaque membra juvant.
>
> *Cloves doe perfume a stincking Breath, and Bind*
> *The Belly: Hence the prime members comfort find.*

And because they are binding (and hot and dry in the third degree) they must not be used, though they help the chiefe parts of Concoction, which are the Stomacke and the Liver, as appeares by the Verses before recited.

The Huskes or Cods of Logwood, or Campeche, are very good, and smell like Fennell; and every one puts in of these, because they are not very hot; though it excuse not the putting in of Annis-seed, as sayes the Author of this Receipt; for there is no *Chocolate* without it, because it is good for many cold diseases, being hot in the third degree; and to temper the coldnesse of the *Cacao*; and that it may appeare, it helpes the indisposition of Cold parts, I will cite the Verses of one curious in this Art:

> Morbosus renes, vesicam, guttura, vulnam,
> Intestina, jecur, cumque lyene caput
> Confortat, variisque Anisum subdita morbis
> Membra: istud tantum vim leve semen habet.
>
> *The Reyns, the Bladder, throat, & thing between—*
> *Enatrailes and Liver, with the Head, and spleen*
> *And other Parts, by ** Annis. it are comforted:*
> *So great a vertue's in that little seed.*

The quantity of a Nut of the *Achiote* Ta-asco. is too little to colour the quantity made according to his Receipt; and therefore, he that makes it, may put in it, as much as he thinkes fit.

Those, who adde Almons, and Nuts, doe not ill; because they give it more body and substance then *Maiz* or *Paniso*A graine like Millet., which others use; and for my part, I should always put it into *Chocolate*, for Almonds (besides what I have said of them before) are moderately hot, and have a thinne juice; but you must not use new Almons, as a learned Author sayes in these Verses.

> Dat modice calidum dulcisque Amigdala succum,
> Et tenuem; inducunt plurima damna nova.
>
> *New Almonds yeild a Hot and slender juice,*
> *But bring new mischiefs by too often use.*

And the small Nuts are not ill for our purpose; for they have almost the temper, which the Almons have; onely because they are dryer, they come nearer the temper of Choler; and doe therefore strengthen the Belly, and the Stomacke, being dryed: for so they must be used for the Confection; and they preserve the head from those vapours, which rise from the Belly: as it appeares by the said Author in these Verses.

> Bilis Avellanam sequitur; sed roborat alvum
> Ventris, & a fumis liberat assa caput.
>
> *Filberds breed Chollar, Th' Belly Fortifie,*
> *Benzoin the Head frees from Fumosity.*

And therefore they are proper for such as are troubled with ventuosities, and *Hypochondriacall* vapours, which offend the brain, and there cause such troublesome dreames, and sad imaginations.

Those who mixe *Maiz* or *Paniso* in the *Chocolate* doe very ill; because those graines doe beget a very melancholly humour: as the same Author expresseth in these Verses.

> Crassa melancholicum præstant tibi Panica succum
> Siccant, si penas membra, gelantque foris.
>
> *Grosse Eares of Corne have Cholorique juice (no doubt)*
> *Which dries, if taken inward; cooles without.*

It is also apparantly windy; and those which mixe it in this *Confection*, doe it onely for their profit, by encreasing the quantity of the *Chocolate*; because every *Fanega* or measure of ** Maiz, or Indian Wheat *Grani* containing about a Bushell and a halfe, is sold for eight shillings, and they sell this *Confection* for foure shillings a pound, which is the ordinary price of the *Chocolate*.

The *Cinamon* is hot and dry in the third degree; it provokes Urine, and helps the Kidneys and Reynes of those who are troubled with cold diseases; and it is good for the eyes; and in effect, it is cordiall; as appeares by the Author of these Verses.

> Commoda & urinæ Cinnamomum, & renibus
> Lumina clarificat, dira venena fugat. (affert:

Cinnamon helps the Reines and Urine well,
It cleares the Eyes, and Poison doth expell.

The *Achiote* hath a piercing attenuating quality, as appeareth by the common practice of the Physitians in the *Indies*, experienced daily in the effects of it, who doe give it to their Patients, to cut, and attenuate the grosse humours, which doe cause shortnesse of breath, and stopping of urine; and so it may be used for any kind of Opilations; for we give it for the stoppings, which are in the breast, or in the Region of the belly, or any other part of the Body.

And concerning the long red Peper, there are foure sorts of it. One is called *Chilchotes*: the other very little, which they call *Chilterpin*; and these two kinds, are very quicke and biting. The other two are called *Tonalchiles*, and these are moderately hot; for they are eaten with bread, as they eate other fruits, & they are of a yellow colour; and they grow onely about the Townes, which are in, and adjoyning to the Lake of *Mexico*. The other Pepper is called *Chilpaclagua*, which hath a broad huske, and this is not so biting as the first; nor so gentle as the last, and is that, which is usually put into the *Chocolate*.

There are also other ingredients, which are used in this *Confection*. One called *Mechasuchil*; and another which they call *Vinecaxtli*, which in the *Spanish* they call *Orejuelas*, which are sweet smelling Flowers, Aromaticall and hot. And the *Mechasuchil* hath a Purgative quality; for in the *Indies* they make a purging portion of it. In stead of this, in *Spaine* they put into the *Confection*, powder of *Alexandria*, for opening the Belly.

I have spoken of all these Ingredients, that every one may make choise of those which please him best, or are most proper for infirmities.

The Second Point.

As concerning the second point, I say, as I have said before, that though it be true, that the *Cacao* is mingled with all these Ingredients, which are hot; yet there is to be a greater quantity of *Cacao*, then of all the rest of the Ingredients, which serve to temper the coldnesse of the *Cacao*: Just as when we seek, of two Medicines of contrary qualities, to compound one, which shall be of a moderate temper: In the same manner doth result the same action and re-action of the cold parts of the *Cacao*, and of the hot parts of the other ingredients, which makes the *Chocolate* of so moderate a quality, that it differs very little from a mediocrity; and when there is not put in any ordinary pepper, or Cloves, but onely a little Annisseed (as I shall shew hereafter) we may boldly say, that it is very temperate. And this may be proved by reason, and experience: (supposing that which *Gallen* sayes, to be true, that every mixt Medicine, warmeth the cold, and cooleth the hot; bringing the examples of Oyle of *Roses*.) By experience, I say, that in the *Indies* (as is the custom of that countrey) I comming in a heat to visite a sick person, and asking water to refresh me, they perswaded mee to take a Draught of *Chocolate*; which quencht my thirst: & in the morning (if I took it fasting) it did warme and comfort my stomack. Now let us prove it by reason. Wee have already proved, that all the parts of the *Cacao* are not cold. For we have made it appeare that the unctuous parts, which are many, be all hot, or temperate: then, though it be true, that the quantity of the *Cacao* is greater than of all the rest of the ingredients, yet the cold parts are at the most, not halfe so many as the hot; and if for all this they should be more, yet by stirring, & mangling of the warme unctuous parts, they are much qualified. And, on the other side, it being mixt with the other Ingredients, which are hot in the second and third degree, being the predominant quality, it must needs be brought to a mediocrity. Like as two men, who shake hands, the one being hot, and the other cold, the one hand borrows heat, and the other is made colder; and in conclusion, neither hand retaines the cold, or heat it had before, but both of them remain more temperate. So like-wise two men, who go to wrestle, at the first they are in their full vigour and strength; but after they have strugled a while, their force lessens by degrees, till at last they are both much weaker, than when they began to wrestle. And *Aristotle* was also of this opinion in his fourth Booke of the Nature of Beasts, *cap. 3*. Where he sayes, that every Agent suffers with the patient; as that which cuts, is made dul by the thing it cuts; that which warmes, cooles it selfe; and that which thrusts, or forceth forward, is in some sort driven bake it selfe.

From whence I gather, that it is better to use *Chocolate*, after it hath beene made some time, a Moneth at the least. I believe this time to be necessary, for breaking the contrary qualities of the severall Ingredients, and to bring the Drinke to a moderate temper. For, as it alwayes falls out at the first, that every contrary will have its predominancy, and will worke his owne effects, Nature not liking well to be heated and cooled, at the same time. And this is the cause why *Gallen* in his twelfth Booke of *Method*, doth advise not to use *Philonium*, till after a yeare, or, at the least, six moneths; because it is a composition made of *Opium* (which is cold in the fourth degree) and of Pepper, and other Ingredients, which are hot in the third degree. This Theorum, and Doctrine, is made good by the practise, which some have made, of whom I have asked, what *Chocolate* did best agree with them? and they have affirmed, that the best is that which hath beene made some moneths: and that the new doth hurt by loosening the Stomack; And, in my opinion, the reason of it is, that the unctuous or fat parts, are not altogether corrected, by the earthy parts of the *Cacao*. And this I shall thus prove; for, as I shall declare hereafter, if you make the *Chocolate* boyle, when you drinke it, the boyling of it divides that fat and oyly part; and that makes a relaxation in the Stomacke in the old *Chocolate*, as well as if it were new.

So that I conclude in this second point, that the *Chocolaticall Confection* is not so cold as the *Cacao*, nor so hot as the rest of the Ingredients; but there results from the action and re-action of these Ingredients, a moderate temper which may be good, both for the cold and hot stomacks, being taken moderately, as shall be declared hereafter; and it having beene made a moneth at the least; as is already proved. And so I know not why any many having made experience of this *Confection* (which is composed, as it ought to be, for every particular) should speake ill of it. Besides, where it is so much used, the most, if not all, as well in the *Indies*, as in *Spain*, finde, it agreeth well with them. He of *Merchena* had no ground in saying, that it did cause Opilations. For, if it were so, the Liver being obstructed, it would extenuate its subject; and by experience, we see to the contrary, that it makes fat; the reason whereof I shall shew hereafter. And this shall suffice for the second Point.

The Third Point.

Having treated in the first poynt, of the definition of *Chocolate*, the quality of the *Cacao*, and of the other Ingredients; and in the second Point, of the Complexion, which results from the mixture of them; There remaines now in the third poynt, to shew the way how to mingle them: And first I will bring the best Receipt, and the most to the purpose, that I could find out; although it be true which I have said, that one Receipt cannot be given, which shall be proper for all; that is to be understood of those, who are sick; for those that are strong, and in health, this may serve: and for the other (as I have said in the conclusion of the first Poynt) every one may make choyse of the Ingredients, as they may be usefull, to this, or that part of his body.

The Receipt is this.

To every 100. *Cacaos*, you must put two cods of the*Chiles long red Pepper, of which I have spoken before, and are called in the *Indian* Tongue, *Chilparlagua*; and in stead of those of the *Indies*, you may take those of *Spaine* which are broadest, & least hot. One handfull of Annis-seed *Orejuelas*, which are otherwise called *Pinacaxlidos*: and two of the flowers, called *Mechasuchil*, if the Belly be bound. But in stead of this, in *Spaine*, we put in six Roses of *Alexandria* beat to Powder: One Cod of *Campeche*, or Logwood: Two Drams of Cinamon; Almons, and Hasle-Nuts, of each one Dozen: Of white Sugar, halfe a pound: of *Achiote* enough to give it the colour. And if you cannot have those things, which come from the *Indies*, you may make it with the rest.

The way of Compounding.

The *Cacao*, and the other Ingredients must be beaten in a Morter of Stone, or ground upon a broad stone, which the *Indians* call *Metate*, and is onely made for that use: But the first thing that is to be done, is to dry the Ingredients, all except the *Achiote*; with care that they may be beaten to powder, keeping them still in stirring, that they be not burnt, or become black; and if they be over-dried, they will be bitter, and lose their vertue. The Cinamon, and the long red Pepper are to be first beaten, with the Annisseed; and then beate the *Cacao*, which you must beate by a little and little, till it be all powdred; and sometimes turne it round in the beating, that it may mix the better: And every one of these Ingredients, must be beaten by it selfe; and then put all the Ingredients into the Vessell, where the *Cacao* is; which you must stirre together with a spoone; and then take out that Paste, and put it into the Morter, under which you must lay a little fire, after the *Confection* is made. But you must be very carefull, not to put more fire, than will warme it, that the unctuous part doe not dry away. And you must also take care, to put in the *Achiote* in the beating; that it may the better take the colour. You must Searse all the Ingredients, but onely the *Cacao*; and if you take the shell from the *Cacao*, it is the better; and when you shall find it to be well beaten, & incorporated (which you shall know by the shortness of it) then with a spoone take up some of the Paste, which will be almost liquid; and so either make it into Tablets; or put it into Boxes; and when it is cold it will be hard. To make the Tablets you must put a spoonfull of the Paste upon a piece of paper, the *Indians* put it upon the leaf of a *Planten-tree*; where, being put into the shade, it growes hard; and then bowing the paper, the Tablet falls off, by reason of the fatnesse of the paste. But if you put it into any thing of earth, or wood, it sticks fast, and will not come off, but with scraping, or breaking. In the *Indies* they take it two severall waies: the one, being the common way, is to take it hot, with *Atolle*, which was the Drinke of Ancient *Indians* (the *Indians* call *Atolle* pappe, made of the flower of *Maiz*, and so they mingle it with the *Chocolate*, and that the *Atolle* may be more wholesome, they take off the Husks of the *Maiz*, which is windy, and melancholy; and so there remaines onely the best and most substantiall part.) Now, to returne to the matter, I say, that the other Moderne drinke, which the Spaniards use so much, is of two sorts. The one is, that the *Chocolate*, being dissolved with cold water, & the scumme taken off, and put into another Vessell, the remainder is put upon the fire, with Sugar; and when it is warme, then powre it upon the Scumme you tooke off before, and so drinke it. The other is to warme the water; and then, when you have put it into a pot, or dish, as much *Chocolate* as you thinke fit, put in a little of the warme water, and then grinde it well with the molinet; and when it is well ground, put the rest of the warme water to it; and so drinke it with Sugar.

Besides these former wayes, there is one other way; which is, put the *Chocolate* into a pipkin, with a little water; and let it boyle well, till it be dissolved; and then put in sufficient water and Sugar, according to the quantity of the *Chocolate*; and then boyle it againe, untill there comes an oyly scumme upon it; and then drinke it. But if you put too much fire, it will runne over, and spoyle. But, in my opinion, this last way is not so wholsome, though it pleaseth the pallate better; because, when the Oily is divided from the earthy part, which remaines at the bottome, it causeth Melancholy; and the oily part loosens the stomacke, and takes away the appetite: There is another way to drink *Chocolate*, which is cold; and it takes its name from the principall Ingredient, and is called *Cacao*; which they use at feasts, to refresh themselves; and it is made after this manner. The *Chocolate* being dissolved in water with the *Molinet*, take off the scumme or crassy part, which riseth in greater quantity, when the *Cacao* is older, and more putrified. The scumme is laid aside by it selfe in a little dish; and then put sugar into that part, from whence you tooke the scumme; and powre it from on high into the scumme; and so drink it cold. And this drink is so cold, that it agreeth not with all mens stomacks; for by experience we find the hurt it doth, by causing paines in the stomacke, and especially to Women. I could deliver the reason of it; but I avoid it, because I will not be tedious, some use it, &c.

There is another way to drinke it cold, which is called *Cacao Penoli*; and it is done, by adding to the same *Chocolate* (having made the *Confection*, as is before set downe) so much *Maiz*, dryed, and well ground, and taken from the Huske, and then well mingled in the Morter, with the *Chocolate*, it falls all into flowre, or dust; & so these things being mingled, as is said before, there riseth the Scum; and so you take and drink it, as before.

There is another way, which is a shorter and quicker way of making it, for men of businesse, who cannot stay long about it; and it is more wholsome; and it is that, which I use. That is, first to set some water to warm; and while it warms, you throw a Tablet, or some *Chocolate*, scraped, and mingled with sugar, into a little Cup; and when the water is hot, you powre the water to the *Chocolate*, and then dissolve it with the Molinet; and then without taking off the scum, drink it as is before directed.

The Fourth Part.

There remaines to be handled in the last Point, of the Quantity, which is to be drunke: at what Time; and by what persons: because if it be drunk beyond measure, not onely of *Chocolate*, but of all other drinkes, or meates, though of themselves they are good and wholsome, they may be hurtfull. And if any finde it Opilative, it comes by the too much use of it; as when one drinkes over much Wine, in stead of comforting, and warming himselfe, he breeds, and nourisheth cold diseases; because Nature cannot overcome it, nor turne so great a quantity into good nourishment. So he that drinkes much *Chocolate*, which hath fat parts, cannot make distribution of so great a quantity to all the parts; and that part which remaines in the slender veines of the Liver, must needs cause Opilations, and Obstructions.

To avoid this inconvenience; you must onely take five or six ounces, in the morning, if it be in winter; and if the party who takes it, be Cholerick, in stead of ordinary water, let him take the distilled water of Endive. The same reason serves in Summer, for those, who take it physically, having the Liver hot and obstructed. If his Liver be cold and obstructed, then to use the water of *Rubarb*. And to conclude, you may take it till the Moneth of *May*, especially in temperate dayes. But I doe not approve, that in the Dogdayes it should be taken in *Spaine*, unlesse it be one, who by custome of taking it, receives no prejudice by it. And if he be of a hot Constitution, and that he have neede to take it in that season, let it, as is said before, be mingled with water of *Endive*; and once in foure dayes, and chiefely when he findes his stomacke in the morning to be weake and fainting. And though it be true, that, in the *Indies*, they use it all the yeare long, it being a very hot Countrey, and so it may seeme by the same reason it may be taken in *Spaine*. First, I say, that Custome may allow it: Secondly, that as there is an extraordinary proportion of heate, so there is also of moisture; which helpes, with the exorbitant heat, to open the pores; and so dissipates, and impoverisheth our substance, or naturall vigor: by reason whereof, not only in the morning, but at any time of the day, they use it without prejudice. And this is most true, that the excessive heate of the Country, drawes out the naturall heate, and disperseth that of the stomack and of the inward parts: Insomuch that though the weather be never so hot, yet the stomack being cold, it usually doth good. I do not onely say this of the *Chocolate*, which, as I have proved, hath a moderate heate; But if you drinke pure wine, be the weather never so hot, it hurts not, but rather comforts the stomack; and if in hot weather you drinke water, the hurt it doth is apparant, in that it cooles the stomack too much; from whence comes a viciated Concoction, and a thousand other inconveniences.

You must also observe, that it being granted, as I have said, that there are earthy parts in the *Cacao*, which fall to the bottome of the Cup, when you make the drinke, divers are of the opinion, that, that which remaines, is the best and the more substantiall; and they hurt themselves not a litle, by drinking of it. For besides, that it is an earthy substance, thick, and stopping, it is of a malancholy Nature; and therefore you must avoid the drinking of it, contenting your selfe with the best, which is the most substantiall.

Last of all, there rests one difficulty to be resolved, formerly poynted at; namely, what is the cause, why *Chocolate* makes most of them that drinke it, fat. For considering that all of the Ingredients, except the *Cacao*, do rather extenuate, than make fat, because they are hot and dry in the third degree. For we have already said, that the qualities which do predominate in *Cacao*, are cold, and dry; which are very unfit to adde any substance to the body. Neverthelesse, I say, that the many unctuous parts, which I have proved to be in the *Cacao*, are those, which pinguifie, and make fat; and the hotter ingredients of this Composition, serve for a guide, or vehicall, to passe to the Liver, and the other parts, untill they come to the fleshy parts; and there finding a like substance, which is hot and moyst, as is the unctuous part, converting it selfe into the same substance, it doth augment and pinguifie. Much more might be said from the ground of Philosophy, and Physique; but because that is fitter for the Schooles, than for this discourse; I leave it, and onely give this Caution, that in my Receipt, you may adde Mellon seeds, and seeds of Pompions of *Valencia*, dryed, and beaten into powder, where there is any heat of the Liver or Kidnyes. And if there be any obstructions of the Liver, or Spleene, with any cold distemper, you may mix the powder of *Ceterach*; to which you may adde Amber, or Muske, to please the scent.

And it will be no small matter, to have pleased all, with this Discourse.

FINIS.